GARÜDIA:

GARIFUNA VISUAL DICTIONARY OF ANIMALS AND FRUITS

(GARIFUNA – ENGLISH – SPANISH)

BY BEN FLORES

First Edition
Copyright 2014 Ben Flores
All rights reserved.

Cover Design: Jason Flores

TABLE OF CONTENTS

DEDICATION

This book is dedicated to my mom and dad, Cecilia and Justin B. Flores, my wife Julia, sons Jason and Kamali and my daughter Kalina Flores.

ACKNOWLEDGEMENTS

First and foremost I would like to give thanks to God Almighty for giving me the wisdom and strength to complete this project. Next I give thanks to my wife, **Julia P. Flores** for putting up with me for all these years, while I was working on this manuscript. Without her patience, this would not have been possible. Also to my two sons, **Jason** and **Kamali Flores**, a huge "seremein" for all the help and patience you have given me, especially during the difficult times I've experienced recently. To my lovely daughter, **Kalina Flores**, seremein for your help and I'm especially proud of you for continuing to pursue your education to earn your Masters in Education.

To my deceased mom and dad, **Cecilia** and **Don Justo Flores,** thank you for nurturing me to become the person that I am. I may not be perfect, but I do the best that I can.

To my beloved sister, **Eva Flores**, and her children, **Lisa, Steff, Charlene** and **Neil**, seremein for being there for your girls and son, and for myself. You have made mom and dad proud. Also, I want to thank my grandchildren **Morgan Sands, Jason Flores Jr, Jayla, Elijah** and **Addison Piper**, and to encourage them to stay in school and get that education, because without it we are nothing. To all my beautiful nieces and nephews, all of whom are unknowingly inspirational at one time or another, I bid all of you a great big "seremein". To **Myisha Battey** and **Nakia Flores**, "Seremein". To my brother, **Joe**, who left us too early, sister **Denise**, nephew **Joe Jr**. and nieces **Rashida** and **Janice Marie Jupiter**, seremein nian hun. Also to my brothers, **Justo, Simeon, Pete and Liston**, a big heartfelt thanks for your help. To my nieces, **Cecilia, Lynden, Briannica** and the others, too numerous to name, Seremein. I want to say seremein to **Clara Flores** and her daughters.

To **Mr. Buck Ciego**, seremein for allowing me and Kamali to be a part of Rootz Novelties. I'm hoping that great things will happen for us this year.

Mr. **Terrence Martinez**, seremein for giving me permission to use your pictures in my book. I also appreciate the kind words of encouragement to complete and publish this labor of isieni (love). Thanks to you, I can now see the light at the end of the tunnel.

Seremein Weyei **Clifford Palacio** for giving me the inspiration to write this book. Your never-ending passion for teaching the Garifuna language to anyone who showed any kind of interest is amazing. I can honestly say that I have learnt a great deal from you. Ouwenbun seremein nichigun bun.

Thank you **Ruben Reyes,** Garifuna scholar and educator born in Tela, Honduras, who edited and published the first *Garifuna Trilingual Dictionary*, available in print and app for mobile devices. Mr. Reyes teaches Garifuna language classes in Los Angeles and is also lead actor, writer, producer, and director of the movie "Garifuna in Peril", available on DVD and digital download worldwide. Ruben is also the inventor of the Garifuna clock, designer of the Garifuna flag emblem, translated the National Anthems of Honduras, Guatemala and the United States into Garifuna, and co-founded the Garifuna Museum of Los Angeles, Garifuna Hope Foundation, Inc. and Garifuna Nation, Inc.

Also to **Carlos Domingo Alvarez,** better known as Mingo, who hails from Trujillo, Colon, Honduras, was a great source of information as far as the Üdüraü "fish" section is concerned. He himself is a fisherman and has a vast knowledge of the different types of fish and their names in Garifuna. Without Mingo's help, this book would have been very difficult to complete. Seremein Mingo for your contribution to this project.

Mr. E. Roy Cayetano, a heartfelt thanks for your contribution to this book. *The People's Garifuna Dictionary* was very instrumental in providing lots of information for my book. I also want to thank you for responding to my many messages on Facebook in a timely manner. Seremein nian bun, ibiri.

Surusia (Dr.) Jorge Bernardez, thank you for being there with your strength and encouragement in creating this project. You might not have been there in flesh, but you were surely there in spirit. Ouwenbu seremein nichigun bun, Nibiri wau.

Alí Allié, thank you for designing and typesetting this book. Alí is the writer, producer and director of two award-winning Garifuna feature films, "El Espíritu de mi Mamá" (Spirit of my Mother) and "Garifuna in Peril".

To **Ms. Adeline "Adz" Valentine** seremein for the picture of your coconut "bagubagutu", in my book. Your timing was priceless. Seremein J., for being there.

Last but not least I thank you, the readers, for your interest in learning Garifuna. Please know that I welcome all constructive criticism about this book. I'm quite sure that there will be certain things that you might not agree with, and that's perfectly okay. If you see anything that you do not agree with, please email me at bendaflo@yahoo.com and I will gladly take your comments under advisement and incorporate them into the next edition.

BDF

INTRODUCTION

by Michele Goldwasser

In writing this book, Ben Flores is continuing a family tradition. His father, Don Justo Flores, sought to preserve the Garifuna language, history, and culture through writing. Don Justo's book entitled *The Garifuna Story, Now and Then* was one of the first works to relate this history from a Garifuna perspective. Perhaps more important, Don Justo Flores worked towards preserving the Garifuna language. He wrote a number of books documenting and teaching the language. He also recognized the need to read and write in Garifuna in order to revitalize the language for generations to come, and he authored books to promote the reading and writing of his language.

In this work, Ben Flores is following in his father's footsteps. This book not only documents images of fauna found in Garifuna communities, it also records the Garifuna terms. This is the first book that focuses on retrieving and documenting these Garifuna words. This is not a simple task. The Garifuna language, like so many other languages throughout the world, is threatened with extinction. While many speakers of Garifuna use the language daily, the possibility that the language will vanish in the next few generations is real. Already, specific words are being lost. This happens for a variety of reasons, including modernization and migration. These transform everyday activities, altering everything from food preparation to work traditions. Fishing and farming, once vital to Garifuna life, are diminishing in importance. The words associated with fishing and farming are also disappearing. Moreover, the flora and fauna of nature are increasingly being relegated to preserves. Words for many species of flora of fauna, no longer encountered on a regular basis, are being forgotten. Perhaps more damaging, English and Spanish words are being substituted for the Garifuna language. Ben Flores has actively conducted research to uncover these Garifuna terms from the collective memories of his community.

Language shift, from Garifuna to English or Spanish, and language loss result not only in the disappearance of words but also in the loss of knowledge. Words in a language relate a culture's history and describe a culture's environment. For example, the number of words listed in the section "Fish-Údüraü" demonstrates two facts. All Garifuna villages were established along coasts, and different varieties of fish were plentiful. In addition, the presence of multiple words in this section illustrates the importance of fishing for the Garifuna people. Furthermore, the use of multiple words for any one species indicates a culture's depth of knowledge of that species. Through the use of words, a culture makes distinctions about the different varieties within any one species expressing their indigenous knowledge of that species. In some cases, scientists have found that indigenous knowledge of flora and fauna exceeds their own understanding of nature. This information exists in language, waiting to be examined by future generations. Without books like this one and efforts by people like Ben Flores, this knowledge could be lost forever.

Most important, language is the voice of the ancestors. Through language, ancestral knowledge passes from one generation to the next. With this book, Ben Flores is helping to preserve these ancestral voices for all to hear. This will help ensure that the Garifuna language and culture will continue for future generations.

ABBREVIATIONS

(Atch)	Attached
(Byr)	Byrsonima
(Dom)	Domesticated
(Fw)	Fresh water
(Jam)	Jamaican
(Ln)	Local name
(Lg)	Large
(Lw)	Low water
(Sw)	Salt water
(Wr)	Wild rabbit

GARÜDIA:

GARIFUNA VISUAL DICTIONARY
OF
ANIMALS AND FRUITS

AMPHIBIANS - BIAN TAWARAGUN - ANFIBIOS

Amphibians are creatures that are cold-blooded vertebrate of the class Amphibia, typically living on land but breeding in water; most lay eggs, but some give birth to live young. Their aquatic larvae (tadpoles) undergo mctamorphosis into the adult form. The class includes the newts and salamanders, frogs, toads, and caecilians.

Bian tawaragun animalu ha awaragubaña dúnarugu luma múarugu.

Anfibios son los animales que pueden respirar tanto dentro del agua como en la tierra.

FROG   HÚA   SAPO

ENGLISH	GARIFUNA	SPANISH	PICTURE
Frog	Húa	Rana	
Toad	Húa	Sapo	

BIRDS - DUNURU - AVES

Birds are any warm-blooded vertebrate of the class Aves, having a body covered with feathers, forelimbs modified into wings, scaly legs, a beak, and no teeth, and bearing young in a hard-shelled egg. The majority of birds are capable of flying.

Dunuru Animalu ha bachati hata, gábutiñu, anihen iyu áhamahagülei hawagu, máritiñu ani ameilihóuwarügütiñu tidan gañé lun háfugachun. Áhamahatiñu híbeagei dunuru.

Aves son animales vertebrados que tienen el cuerpo cubierto de plumas, tienen pico, alas y nacen de huevo. La mayoría de las aves pueden volar.

MACAW   GARARAWA   GUACAMAYO

ENGLISH	GARIFUNA	SPANISH	PICTURE
Blackbird	Würidiri; Würindiri	Mirlo	
Booby (blue footed)	Barahawa (dingu tugudi)	Piquero patiazul	
Nightjar; Bullbat (ln)	Duguyu (Huyu)	Chotacabras	
Chachalaca (plain) (ln)	Chagalaga; Hádaraga	Chachalaca	
Coot	Fuliku	Fochas	

ENGLISH	GARIFUNA	SPANISH	PICTURE
Crane	Yábura	Grulla	
Crested Guan	Fawa gábuluguse	Pava moñuda	
Crow	Wuridiu	Cuervo	
Curassow (ln) (Wild turkey)	Ugu	Paujil	

ENGLISH	GARIFUNA	SPANISH	PICTURE
Dove	Wagugua	Paloma	
Drake	Ganaru (wügüri)	Pato doméstico	
Duck	Ganaru (würi)	Pato doméstico	
Eagle	Habisa gayu	Águila	

ENGLISH	GARIFUNA	SPANISH	PICTURE
Egret	Gálini	Garceta	
Falcon	Garún	Halcón	
Frigate bird	Moure	Pájaro de Fragata	
Gallinule	Yábura	Garza	

ENGLISH	GARIFUNA	SPANISH	PICTURE
Gannet (red-footed)	Barahawa (funa tugudi)	Alcatraz patirrojo	
Golden Oriole	Gurasue; Yágüri	Oropéndola	
Goose	Wánana	Ganso	
Grackle	Sanadi	Zanate	
Grebe Diving Dipper (ln)	Bulú-bulu sega	Colimbo	

ENGLISH	GARIFUNA	SPANISH	PICTURE
Hawk	Garún	Halcón	
Hen	Ariran (wuri); Gayu; Yúgurududu	Gallina	
Hen-Guinea	Giniheni	Gallina guinea	
Heron	Hamamali	Garza	

ENGLISH	GARIFUNA	SPANISH	PICTURE
Humming bird	Yürüdü	Colibrí	
Jakana Yellow tail (ln)	Yagei	Jacanas	
Jay bird	Piánpiamu	Arrendajo	
Kingfisher	Sada	Martín pescador	
Macaw	Gararawa	Guacamayo	

ENGLISH	GARIFUNA	SPANISH	PICTURE
Mockingbird	Solan	Sinsonte; Zenzontle	
Mourning dove	Duguruwe	Paloma torcaza	
Owl	Gáduri	Búho	
Parakeet	Gareigarei	Perico	

ENGLISH	GARIFUNA	SPANISH	PICTURE
Parrot	Gurewegi	Loro	
Partridge	Féindanu	Perdiz	
Peacock	Fawa	Pavo real	
Pelican	Yalifu	Pelícano	
Pigeon	Unhun	Paloma	

ENGLISH	GARIFUNA	SPANISH	PICTURE
Quail	Kueri	Codorniz	
Rooster	Ariran (wuguri)	Gallo	
Sandpiper; Snipe	Diu	Andarríos	
Seagull	Sáwarawára	Gaviota	
Sparrow	Haradoudou	Gorrión; Passeridae	

ENGLISH	GARIFUNA	SPANISH	PICTURE
Stork	Sigueyan	Cigüeña	
Swallow	Wéigaraga	Golondrina	
Toucan	Bilibudu	Tucán	
Turkey	Figaga; Dururunbe	Pavo; Guajolote	

ENGLISH	GARIFUNA	SPANISH	PICTURE
Vulture; (John Crow) (ln)	Wadubi	Buitre; (Zopilote)	
Woodpecker	Cheku; Dúnduwewe	Pájaro carpintero	
Yellow tail (ln)	Yagei; Yágüri	Tejedor	

CRUSTACEANS - WAYUMUNA - CRUSTÁCEOS

Crustaceans any chiefly aquatic arthropod, of the class Crustacea, that live mostly in water, typically having a segmented body covered with a hard shell or crust and jointed appendages. These include: lobsters, shrimps, crabs, barnacles, copepods and wood lice.

Wayumuna hagía aganóubaña dúnarugu lidan sun luruyeri luliligan, dandu lidan barana, keisi dúnarugu; deretu hagei, asaraguatu hagudi ani. Añahán besafu hádangien: húgawaru, isuru, hürü, bagarabudu, herenge, bawasa, hama ámuraüya.

Los Crustáceos son fundamentalmente acuáticos y habitan en todas las profundidades, tanto en el medio marino, salobre y de agua dulce; unos pocos han colonizado el medio terrestre, como la cochinilla de la humedad (isópodos).

LOBSTER   HÚGAWARU GÁRITI   BOGAVANTE

ENGLISH	GARIFUNA	SPANISH	PICTURE
Crab (beach)	Herenge; Heringe	Cangrejo Fantasma	
Crab (blue)	Hamaba; Harouru; Hidururu	Jaiba	
Crab (fiddler)	Bágárabúdu; Bágáragabúda	Cangrejo Violinista	
Crab (hairy)	Gusa	Cangrejo peludo	
Crab (hermit)	Magüre	Cangrejo ermitaño	
Crab (land)	Bágübagü; Holbochi; Hürü; Wayumu	Cangrejo	

ENGLISH	GARIFUNA	SPANISH	PICTURE
Crab (red)	Bawasa; Funerewa	Cangrejo rojo	
Crab (sand)	Leweyu	Chiquilique; Ligia oceánica; Cangrego de arena	
Crab (soldier)	Hawayule	Cangrejo soldado	
Crawfish (fw)	Isuru deré tagei	Cangrejos de río	
Lobster	Húgawaru gáriti	Bogavante	

ENGLISH	GARIFUNA	SPANISH	PICTURE
Lobster (spiney)	Húgawaru	Langosta	
Shrimp (fw)	Isuru	Chacalín	
Shrimp (sw)	Isuru; Gamaroun	Camarón	

FISH - ÚDURAÜ - PEZ

Fish are any of various cold-blooded, aquatic vertebrates, characteristically having gills, commonly fins, and typically an elongated, streamlined body covered with scales.

Úduraü sun animalu baranahana luma dúnaruguna ha dilí-tita, gábutiñu, gayamaratiñu, labu arünaü lánina áfulihani. Migifeguati hágubu, bürüraütiñu luma ti hiyeróuti lá hara.

Los peces son animales vertebrados acuáticos, generalmente ectotérmicos, la mayoría de ellos recubiertos por escamas, y dotados de aletas, que permiten su desplazamiento en los medios acuáticos, y branquias, con las que captan el oxígeno disuelto en el agua.

JACK   GULILAWARU   JURELES

ENGLISH	GARIFUNA	SPANISH	PICTURE
Angel fish	Pámpana; Síegugu	Angelote	
Barracuda	Yamura	Barracuda	
Bonefish	Masimasi	Macabijo; Macabí	
Bonito	Bunigü	Bonito; Sarda	
Catfish (fw)	Buganchu	Bagre	
Catfish (sw)	Bágüri	Bagre	

ENGLISH	GARIFUNA	SPANISH	PICTURE
Cichlid (Tubah) (ln)	Duba	Trucha	
Cling fish	Fegi	Peces pegados	
Codfish	Lamuri	Bacalao	
Drummer fish (Black Seabream)	Duroma; Gurín; Guhua	Curbina	
Eel (conger)	Muingi; Wurigamara	Anguila; Congrio	
Flounder fish	Fadabelu; Gúnfuligun	Lenguado	
Flying fish	Wadanaru	Pez volador	

ENGLISH	GARIFUNA	SPANISH	PICTURE
Fowl fish	Gayu úduraü	Pez de los Sargazos	
Gray Snapper	Hiyawa	Pargo gris	
Grouper fish	Gurupa; Kurupa	Mero	
Grunt fish	Guéruguru	Roncos	
Grunt fish (French)	Yakuma	Ronco amarillo	
Grunt fish (lw)	Yawanta	Ronco	

ENGLISH	GARIFUNA	SPANISH	PICTURE
Halibut	Badasi	Fletán	
Hammer fish	Gabuteri	Pez martillo	
Jack (big eye) (Horse-eye)	Masouri	Carángido	
Jack fish (largest)	Garudusu	Pez jack (más grande)	
Jack fish	Yawarigaü	Pez Jack	
Jack (Skip)	Gulilawaru	Jureles	

ENGLISH	GARIFUNA	SPANISH	PICTURE
Jew fish	Inegu	Epinephelus itajara	
King fish	Awawi	Scomberomorus cavalla	
Mackerel (king)	Warubi	Macarela	
Mackerel (lg)	Baganúhobu	Verdel	
Manta-ray; Devilfish	Lareimu	Mantarraya	
Marlin	Barawagütü	Pez Espada	
Mudfish (fw)	Gusuna; Égewe	Peces barro	

ENGLISH	GARIFUNA	SPANISH	PICTURE
Mudfish (sw)	Maguba	Peces pulmonados	
Mullet fish	Badasi; Barasi	Salmonete	
Needlefish (Long Guard) (ln)	Góunbiri	Agujeta	
Parrot fish	Dudei	Pez loro	
Porcupine fish	Chufuyuma	Pez puerco espín	
Porgy fish	Gériri	Pargo	
Red Snapper	Hiyawa	Huachinango	

ENGLISH	GARIFUNA	SPANISH	PICTURE
Remora (Sucker fish) (ln)	Fegi	Rémora	
Rockfish	Guchelele	Pez piedra	
Salmon	Sámani	Salmón	
Sardine	Dáurali; Wiwi sasa; Sansua	Sardina común	
Sardine (lg)	Dawarari; Gurisawa	Sardina	
Shad fish (Stone Bass) (ln)	Gawachau	Cawacha; Mero de roca	

ENGLISH	GARIFUNA	SPANISH	PICTURE
Shark	Wéibayuwa	Tiburón	
Shark (Hammer-head)	Bayara; Gabuteri	Pez martillo	
Shark (nurse)	Aleiru	Tiburón nodriza	
Shark (tiger)	Masíradi	Tiburón Tigre	
Shark (lemon)	Wéibayuwa Síndurunouti	Tiburón limón	
Shark (Black tip)	Wéibayuwa Wurí tígiri	Tiburón de puntas negras	

ENGLISH	GARIFUNA	SPANISH	PICTURE
Shark (Saw fish)	Gasiaran	Pez sierra	
Shark (White tip)	Wéibayuwa Haru tígiri	Tiburón de puntas blancas	
Sheephead fish	Sígugu	Sargo chopa	
Snake fish	Dugunuwe	Pez bívora	
Snapper (cubera)	Gagubanagei	Pargo cubera	
Snapper (mangrove)	Gagubanagei	Pargo manglero	

ENGLISH	GARIFUNA	SPANISH	PICTURE
Snapper (mutton)	Gagubanagei	Gambas	
Snapper (silk)	Galali	Pargos; Huachinangos	
Snapper (yellowtail)	Yagiri	Pargo amarillo	
Snook fish	Bürütóuba; Kalúa	Robalo común	
Sprat fish	Gurisawa	Espadín	
Stingray	Sibari	Raya	
Swordfish	Baráu; Barawagütü	Pez espada	

ENGLISH	GARIFUNA	SPANISH	PICTURE
Tarpon	Haba	Sábalo; Tarpón	
Toadfish	Limilia	Pez sapo	
Vieja; Old wife (ln)	Hanóun	Vieja	
Whipray	Wagawa	Chupare Oval	
Whiting	Hachulali	Merlán	
Wrasses creole (ln)	Gériri	Lábridos	

FRUIT - TIN WEWE - FRUTA

Fruits are the usually edible reproductive body of a seed plant, especially one having a sweet pulp associated with the seed. The span of the fruit of the tree is a product of fertilization in a plant with its modified envelopes or appendages specifically: the ripened ovary of a seed plant and its contents.

Tin wewe ligía tagindubei wewe lárigi tegelewehan fuluri, gabimedeitu tíbeagei dan to funaru. Líbeagei dan geiguwatu, gabunúati tabu odi tila lun táhüchün iseri wewe tídangien. Ganügatiñu giñe dunuru lun tin wewe lidoun amu fulasu lun tawaraüdagun. Anuhoun turuyeri tin wewe lídangien tugudina láñagua tídibu.

El fruto es el órgano procedente de la flor, o de partes de ella, que contiene a las semillas hasta que estas maduran y luego contribuye a diseminarlas.

CACOA   GABU   CACOA

ENGLISH	GARIFUNA	SPANISH	PICTURE
Acorn	Malalei	Bellota	
Almond; Hammans	Hamanasi	Almendro	
Annatto	Gusewe	Achiote	
Apple	Ápuli	Manzana	
Avocado	Wagadi	Aguacate	

ENGLISH	GARIFUNA	SPANISH	PICTURE
Baboon cap	Monkikiapu	Ospip	
Banana (apple)	Bágugu	Plátano malayo	
Banana	Bímina; Bímena	Plátano	
Basil	Basein	Albahaca	
Bay Leaf	Warúaru	Hoja de Laurel	

ENGLISH	GARIFUNA	SPANISH	PICTURE
Beans	Eifi	Frijol	
Beans (navy)	Bámbaragu	Frijoles	
Beans (red kidney)	Eifi-funati	Frijoles rojo	
Beans (pinto)	Eifi-singa	Frijol Pinto	
Beans (black)	Eifi-wuriti	Frijoles negro	

ENGLISH	GARIFUNA	SPANISH	PICTURE
Beans (butter)	Magundan	Habas	
Berries (coco-plum) (ln)	Higagu	Hicacos	
Bitter melon; Cerasee (ln)	Sorosí; Papina	Calaica	
Black Pepper	Furefuru	Pimienta negra	
Bread	Fein	Pan	

ENGLISH	GARIFUNA	SPANISH	PICTURE
Breadfruit	Yamichuga; Ayuname	Masapan	
Butter	Beru	Mantequilla	
Cabbage	Repoyu	Repollo; Col	
Carao; Buccut (ln)	Karóu	Carao	
Cassava	Ereba	Cassabe	
Cashew	Uruwi; Uri	Marañón	

ENGLISH	GARIFUNA	SPANISH	PICTURE
Cheese	Furumasu	Queso	
Coco plum	Higagu	Hicacos	
Cocoa	Gabu	Cacao	
Coconut	Fáluma	Cocotero	
Coconut (baby)	Bayubayu	Coco tierno	

ENGLISH	GARIFUNA	SPANISH	PICTURE
Coconut (young)	Guguedi	Coco de agua	
Coconut Oil	Garaba; Tagüle Fáluma	Aceite de coco	
Cohune nut	Yawara	Corozo	
Corn	Awasi	Maíz	
Crabboe (ln); Nance (byr)	Muréi	Nance	

ENGLISH	GARIFUNA	SPANISH	PICTURE
Custard apple	Gasima	Anón	
Egg	Gañe	Huevo	
Garlic	Lai	Ajo	
Grape	Beibei	Uvas	

ENGLISH	GARIFUNA	SPANISH	PICTURE
Grapefruit	Charigi	Toronja	
Guava	Wériafa	Guayaba	
Jack fruit	Panpán	Fruta de jaca	
Lemon	Sínduru	Limón	
Lime	Sínduru	Limón	

ENGLISH	GARIFUNA	SPANISH	PICTURE
Locust; (Stinking toe)	Góubali; Garubali	Copal	
Mamey apple	Sabudi	Zapote	
Mango	Mangu	Mango	
Medicine	Árani	Medicina	
Milk	Miligi	Leche	

ENGLISH	GARIFUNA	SPANISH	PICTURE
Monkey Apple	Uragu	Uraco	
Nutmeg	Misigadu	Nuez moscada	
Okra	Nehu	Okra	
Onion	Sebuya	Cebolla	
Orange	Aransu	Naranja	

ENGLISH	GARIFUNA	SPANISH	PICTURE
Oregano	Oreganu	Orégano	
Pacay	Béirawa; Gurehen; Bri-bri	Guamo	
Papaya	Ababaü	Papaya	
Passion fruit	Le megu; Mereguia	Granadilla	
Peanut	Pindá; Mali; Finadu	Cacahuate; Maní	

ENGLISH	GARIFUNA	SPANISH	PICTURE
Pepper	Ati	Chile Habanero	
Pepper (bell)	Ati bímeti	Chile Dulce	
Pineapple	Yéyewa	Piña	
Plantain	Baruru	Plátano	
Plantain (atch)	Mándaya	Gemelos	

ENGLISH	GARIFUNA	SPANISH	PICTURE
Plantain (Jam)	Chata; Fulagoun	Chatos	
Plum	Siringuela	Ciruela	
Provision fruit	Gumaga	Zapotón; Ceiba	
Pumpkin	Wéiyama	Ayote; Calabaza	

ENGLISH	GARIFUNA	SPANISH	PICTURE
Rice	Ri/Rin	Arroz	
Salt	Salu	Sal	
Sapodilla	Sabadüle	Chicle	
Sorrel	Hameiga	Alazán; Jamaica	
Sour orange	Kahela	Naranja agria	

ENGLISH	GARIFUNA	SPANISH	PICTURE
Soursap	Gurusulu	Guanábana	
Squash	Fisudi wéyema	Calabacera	
Star apple	Starápuli	Caimito	
Star fruit	Karanbola	Carambolas	
Sugarcane	Asigaru; Gániesi	Caña de azucar	

ENGLISH	GARIFUNA	SPANISH	PICTURE
Tamarind	Tánparanu	Tamarindo	
Tomato	Dumadi	Tomate	
Tortilla (flour)	Durudia	Tortilla	
Watermelon	Badía	Sandía	

INSECT - IGEI - INSECTO

Insects are any animal of the class Insecta, comprising small, air-breathing arthropods having the body divided into three parts (head, thorax, and abdomen), and having three pairs of legs and usually two pairs of wings. It's also a small arthropod, such as a spider, tick, or centipede, having a superficial, general similarity to the insects.

Igei hagía sun harahüñü animalu, awaraguatiñu garabali, fareinti hágubu lidan üruwa murusun (ábulugu, aniguagu luma uragei), luma giñe ti anihein la üruwa feru ugudi hawagu, añahein hádangien garünatiñu. Añahein giñe igei gíbetimati ugudi hawagu kéiburi: anasi, gubari hama san-tugudi.

Los insectos, son una clase de animales invertebrados, del filo de los artrópodos, caracterizados por presentar un par de antenas, tres pares de patas y dos pares de alas (que, no obstante, pueden reducirse o faltar).

SCORPION   ÁGURU   ALACRÁN

ENGLISH	GARIFUNA	SPANISH	PICTURE
Ant (fire) (ln)	Budundu; Harü	Hormiga	
Ant (Giant)	Hawasali; Chichawali	Hormiga	
Ant (Parasol) (wee-wee) (ln)	Sarisari	Zompopo	
Bedbug	Chincha	Chinche	
Bee	Maba	Abeja	

ENGLISH	GARIFUNA	SPANISH	PICTURE
Bottleass (ln) (simulid flies)	Mebebegi	Simúlidos	
Butterfly	Waribagabaga; Würigabaga	Mariposa	
Chigoe; Chigger	Sigü	Chigoe	
Cicada	Sídigidigi	Cigarra	
Cricket	Diru; Sídigiri	Grillo	

ENGLISH	GARIFUNA	SPANISH	PICTURE
Dragon fly	Sínganga; Híngigi	Libélula	
Firefly	Wádigidigi; Gemegeme	Luciérnaga	
Flea	Hayaba	Pulga	
Fruitfly	Minimini	Tefrítido	
Grasshopper	Chagagaru	Saltamontes	

ENGLISH	GARIFUNA	SPANISH	PICTURE
Horsefly	Gunga; Mabarawa	Tábanos	
Housefly	Werewere	Mosca	
Locust	Chiganaru	Langosta; Algarroba	
Louse	Iein; Hiñei	Piojo	
Mole Cricket	Chigili-chigili	Grillotalpa; Topogrillo	

ENGLISH	GARIFUNA	SPANISH	PICTURE
Mosquito	Marin	Mosquito; Zancudo	
Mosquito larva	Mábulechu; Gubigubi	Larva de mosquito	
Moth	Lige lampu	Polilla; Heterocera	
Roach; Cockroach	Fudi	Cucaracha	
Sandfly	Mabiri	Jején	

ENGLISH	GARIFUNA	SPANISH	PICTURE
Scorpion	Águru; Sigu	Alacrán	
Spider	Anasi; Gúrenren/Gú wewen	Araña	
Tarantula	Warouri (u)	Tarántula	
Tick	Gubari	Garrapata (Patacón)	

ENGLISH	GARIFUNA	SPANISH	PICTURE
Wasp	Hulahüñü	Avispa	
Weevil	Hibigi; Lige eifi	Gorgojo	
White ant	Hamutiri	Comején	

INVERTIBRATE - MÁBUTIÑU - INVERTEBRADO

Invertebrates are animal species that do not develop a vertebral column. This in effect includes all animals apart from the subphylum Vertebrata. Familiar examples of invertebrates include insects, worms, clams, crabs, octopus, snails, and starfish.

Mábutiñu hagía animalu mábuti hanagan. Habadinagua mábutiñu dariwatiñu igei, héweraü, síndiri, hürü, arari, gureme hama átürübeigan.

Se denomina **invertebrados** a todos aquellos animales (reino Animalia) que no se encuadran dentro del subfilo de los vertebrados del filo cordados.

WOLLY BEAR   SUSU BLÁNGIDU   ORUGA

ENGLISH	GARIFUNA	SPANISH	PICTURE
Centipede	Íliba; Íluba	Ciempiés	
Earthworm	Baridiga; Héwerou	Gusano; Lombriz de tierra	
Jelly fish	Gafiloun; Garungaru; Gougou	Medusa	
Leech	Sóku-soku	Sanguijuela	
Millipede	Íluba	Milpiés	
Wolly Bear	Susu blángidu	Oruga	

MAMMALS - ÁTAHATIÑU - MAMIFERO

Mammals are any vertebrate of the class Mammalia, having the body more or less covered with hair, nourishing the young with milk from the mammary glands, and with the exception of the egg-laying monotremes, giving birth to live young.

Átahatiñu hagía animalu guratiñu uriraü, háyuti hara hibeagei, guragüdatiñu míligi houn harahüñü lídangien uri.

Los mamíferos (mammalia) son una clase de vertebrados amniotas homeotermos (de "sangre caliente"), con pelo y glándulas mamarias productoras de leche con la que alimentan a las crías. La mayoría son vivíparos (con la notable excepción de los monotremas: ornitorrinco y equidnas).

LEOPARD   GÉIGUSI   LEOPARDO

ENGLISH	GARIFUNA	SPANISH	PICTURE
Agouti	Aguri; Inaki	Agutí	
Anteater	Haspia	Oso hormiguero	
Antelope	Antilunpu; Usari	Antílope	
Ape	Sisimidu	Simio	
Armadillo	Gasigamu	Armadillo	
Baboon	Babunu	Babuino	

ENGLISH	GARIFUNA	SPANISH	PICTURE
Bat	Búriri	Murciélago	
Bear	Osu	Oso	
Bull	Habuyaba	Toro	
Cat	Mesu	Gato	
Coati; Quash (ln)	Fisudi	Coatí; Pizote	

ENGLISH	GARIFUNA	SPANISH	PICTURE
Cougar	Mótete	Puma	
Cow	Adabiu; Bágasu (wuri)	Vaca	
Deer	Usari	Ciervo; Venado	
Dog	Ounli; Yúledu; Filuiri	Perro	
Doe	Würi usari Hiañru usari	Cierva	

ENGLISH	GARIFUNA	SPANISH	PICTURE
Dolphin	Sigei; Sigiri	Delfín	
Donkey	Burigü (würi)	Burro	
Elephant	Lefan	Elefante	
Ewe (fem)	Mudún	Oveja	
Ferret	Férudu	Hurón	
Fox	Fágusi	Zorro	

ENGLISH	GARIFUNA	SPANISH	PICTURE
Goat	Gábara	Cabra	
Hare	Másaraga (árabuna)	Liebre	
Hog	Buíruhu; Gúnguti	Cerdo; Chancho	
Horse	Gabayu; Suáfuru	Caballo	
Jackal	Ounli árabuna	Chacal	

ENGLISH	GARIFUNA	SPANISH	PICTURE
Jackass	Burigü (wügüri)	Burro	
Lamb; Sheep	Mudún	Cordero; Oveja	
Leopard	Géigusi	Leopardo	
Lion	Liún (wuguri)	León	
Lioness	Liún (wuri)	Leona	

ENGLISH	GARIFUNA	SPANISH	PICTURE
Manatee	Manadi	Manatí	
Margay	Mesu-géigusi	Gato tigre	
Mermaid	Meru-meru; Agayuma	Sirena	
Mongoose	Habisa hewe	Herpéstido	
Monkey (spider)	Megu anasi	Monos araña	

ENGLISH	GARIFUNA	SPANISH	PICTURE
Monkey (howler)	Megu; Olingu	Monos aulladores	
Mouse	Garadun	Ratón	
Mule	Mile	Mula	
Otter; Water-dog (ln)	Wátadagü; Ounli dúnaruguna	Nutria; Cuica de agua	
Ox	Bágasu (wügüri); (eyeri)	Buey	

ENGLISH	GARIFUNA	SPANISH	PICTURE
Paca; Gibnut (ln)	Gibinadu	Tepeizcuinte	
Pig	Buíruhu; Gúgunti	Cerdo; Chancho; Puerco	
Porcupine	Fadigiliyeru; Fidigiliyeru	Puerco espín	
Possum (ln); Opossum	Daguasi; Manihu; Pásamu	Zarigüeya	
Rabbit (dom)	Másaraga	Conejo	
Racoon	Mapachi; Rágunu	Mapache	

ENGLISH	GARIFUNA	SPANISH	PICTURE
Rat	Garadun	Rata	
Skunk (Polecat) (ln)	Mabiridu	Mofeta; Zorrillo	
Squirrel	Bigiburu	Ardilla	
Stag	Usari	Venado	
Tapir; Mountain cow (ln)	Dandei; Sabadiuwa	Tapir	

ENGLISH	GARIFUNA	SPANISH	PICTURE
Tiger	Deguan; Géigusi	Tigre	
Weasel	Beledu	Comadreja	
Whale	Amana; Udu; Masíradi	Ballena	
Wild boar; Pecarry (ln)	Gegeu; Hawia	Habalí	

MOLLUSKS - ÑULERU - MOLUSCOS

Mollusks are any of the numerous chiefly marine invertebrates of the phylum Mollusca, typically having a soft un-segmented body, a mantle, and a protective calcareous shell and including the edible shellfish and the snails.

Ñuleru hagía hádagien animalu ha gíbetimabaña hadan sun animalu. Mábutiñu, líbeagei dan gáfadagunti hágubu le huáliti, anihéin giñe aban agei hawagu. Añahein wadabu hadan hama giñe guwe.

Los moluscos forman uno de los grandes filos del reino animal. Son invertebrados protóstomos celomados, triblásticos con simetría bilateral (aunque algunos pueden tener una asimetría secundaria) y no segmentados, de cuerpo blando, desnudo o protegido por una concha.

CONCH   WADABU   CARACOL

ENGLISH	GARIFUNA	SPANISH	PICTURE
Clam	Síndiri	Almejas	
Conch	Wadabu	Caracol	
Conch (mud)	Pangu	Pango	
Conch (mule)	Warafiaga; Tugudi mile	Casco de burro	
Conch Shell	Wadabágei	Concha	
Octopus	Arari	Octópoda	

ENGLISH	GARIFUNA	SPANISH	PICTURE
Sea snail	Giua	Caracolito	
Sea star	Waruguma	Asteroideo	
Sea urchin	Águnuru	Erizos de mar	
Sea worm	Bereuera	Gusano de mar	
Shell fish	Guwe	Conchas	
Snail	Gureme	Caracol de jardín	

REPTILES - ADURUGUÁTIÑU - REPTILES

Reptiles are any cold-blooded vertebrate of the class, Reptilia, including tortoises, turtles, snakes, lizards, alligators, crocodiles, and extinct forms.

Aduruguátiñu hagía aban haruyeri animalu ha aduruhatiñu houngua luagu mua, kéiburi bugudura, hewe, wagagan, wayamaga, garangadili, agare, hama ámuraüya.

Los reptiles (Reptilia) son un grupo parafilético de vertebrados amniotas provistos de escamas epidérmicas de queratina. Estos incluye las testudíneos, tortugas marinas, culebras, lagartos, aligatores, cocodrilos e iguanas marinas.

TURTLE   ARAGAGA   TORTUGA

ENGLISH	GARIFUNA	SPANISH	PICTURE
Alligator	Agare	Lagarto	
Caiman	Garangadili	Caimán	
Crocodile	Agareina; Agarewa	Cocodrilo	
Fer-De-Lance	Haua	Crotalino	
Iguana (fem)	Wayamaga	Iguana	
Iguana (m)	Serewei; Gorobu	Garrobo	

ENGLISH	GARIFUNA	SPANISH	PICTURE
Lizard	Wagagan	Lagartija	
Snake Tommy Goff (ln)	Damagasi	Tamagás	
Snake (boa) Wowla (ln)	Üwi	Boa constrictora	
Snake (coral)	Kurali	Serpiente de coral	
Snake (green)	Árubana	Serpiente verde	
Snake; Viper	Hewe	Culebra; Víbora	

ENGLISH	GARIFUNA	SPANISH	PICTURE
Turtle Bocotora (ln)	Aragaga	Tortuga	
Turtle (green)	Gadaru	Tortuga verde	
Turtle (hawksbill)	Gáraru	Carey	
Turtle (hicatee)	Bugudura; Higidi	Jicotea elegante	
Turtle (loggerhead)	Gawamu	Tortuga de mar	
Turtle (swanka)	Bugudura	Tortuga de orejas rojas	

TUBERS - ÍLUREI - TUBÉRCULOS

Tubers are various types of modified plant structures that are enlarged to store nutrients, and are used by plants to survive the winter or dry months. They provide energy and nutrients for regrowth during the next growing season, and reproduce by means of asexual reproduction. There are both stem and root tubers.

Ílurei tuguya aban turuyeri ílagülaü migifeguatu to árügüdütu lisuni sagoun lun takipurunu túmagua, óuseratumutu ídibu lidan lidiliga luma lidan lanarime lárouga. Rutu erei lun taheredun ídibu lidanya lewengua idani. Tabuguarügü tagüriaha turuyeri ídibu to. Anuhoun biama turuyeri ílurei: to surugutu tuma to gálureguárügütu.

Un **tubérculo** es un tallo subterráneo del subsuelo modificado y engrosado donde se acumulan los nutrientes de reserva para la planta. Posee una yema central de forma plana y circular. No posee escamas ni cualquier otra capa de protección, tampoco emite hijuelos.

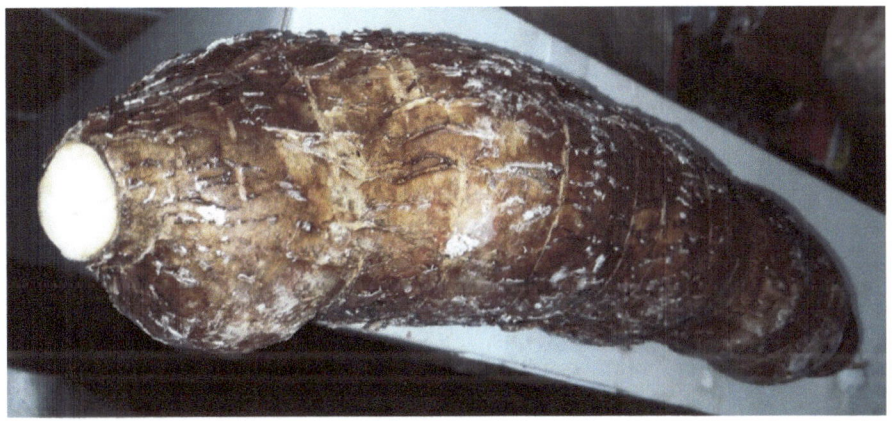

CASSAVA (SWEET)   GUMANANA   YUCA AMARGA

ENGLISH	GARIFUNA	SPANISH	PICTURE
Arrowroot	Inginára	Arrurruz	
Cassava (bitter)	Gain	Yuca Amarga	
Cassava (sweet)	Gumanana	Yuca de comer	
Coco (root)	Wahú	Malanga	
Ginger	Chichanbara	Jengibre	

ENGLISH	GARIFUNA	SPANISH	PICTURE
Jicama	Higama	Jicama	
Potato (Irish)	Mingoun	Papa; Patata	
Potato (sweet)	Mabi; Melewe	Camote	
Yam	Amanu	Batata; Ñame	
Yam (yellow)	Simuruna; Yami	Ñame	
Yampa	Guchu	Ñame	

CITES

Flores, Don Justo. *The Garifuna Story, Now and Then Book One.*

Flores, Don Justo. *A Study in the Reading and Writing of Garifuna.*

Reyes, Ruben. *Garüdia: Garifuna Trilingual Dictionary.*

Cayetano, E. Roy. *The People's Garifuna Dictionary.*

Google.com. Public Domain Images on Google.

Houghton Mifflin Company. *Inside Guides Belize*

ENGLISH WORD ALPHABETICAL INDEX

GARIFUNA WORD ALPHABETICAL INDEX

SPANISH WORD ALPHABETICAL INDEX

ABOUT THE AUTHOR

Ben Flores was an Assistant Producer for the movie "Garifuna In Peril" and was responsible for the subtitles and translations. He also was an Assistant Producer for the short film "Drums" by Simeon Flores.

He is a graduate of National University with a B.S. in Computer Science, is curator of the Garifuna Museum of Los Angeles (GaMoLA) and co-owner of Rootz Novelties along with Buck Pastor & Kamali Flores, a source for Garifuna-themed clothing, crafts and accessories. He is the son of the late Don Justo Flores, a musician, writer, playwright, and machinist whose books include *The Garifuna Story, Now & Then (1) & (2)*, *The History of the Writing of Garifuna in Our Time*, *A Study in the Reading & Writing of Garifuna*, *Now or Never to the Future of the Garifuna Language*, *Bilingual Garifuna Study*, and a few more. Don Justo, (as he was affectionately known) was also a musician who produced many albums: "The Garifuna Singers, Don Justo & Dolores Moralez, with the Ritmo Caribe Combo", "Reggae Nuwanee 1", "The Great Isabel Flores", "Guribiyuwa", T.V. Ramos and finally, a Christmas album entitled "Furieyegi Leremu Fedu".

Presently, Ben is also a member of Garifuna Hope Foundation. *Garifuna Visual Dictionary of Animals* is his first book. "I thought that by writing this book, this would be another way of preserving our precious language."

BDF

NOTES